27th Topical Conference on Refinery Processing 2024

Topical Conference at the 2024 AIChE Spring Meeting and 20th Global Congress on Process Safety

New Orleans, Louisiana, USA
24-28 March 2024

ISBN: 979-8-3313-1709-6

Printed from e-media with permission by:

Curran Associates, Inc.
57 Morehouse Lane
Red Hook, NY 12571

Some format issues inherent in the e-media version may also appear in this print version.

AIChE
120 Wall Street, FL 23
New York, NY 10005-4020

Phone: (800) 242-4363
Fax: (203) 775-5177

www.aiche.org

Additional copies of this publication are available from:

Curran Associates, Inc.
57 Morehouse Lane
Red Hook, NY 12571 USA
Phone: 845-758-0400
Fax: 845-758-2633
Email: curran@proceedings.com
Web: www.proceedings.com

TABLE OF CONTENTS

PLANT SIMULATION AND DESIGN

DIGITIZATION AND CONTROLS

PLANT SIMULATION AND DESIGN II

RENEWABLE FUELS II

139a Aerogel Innovations: Harnessing Reduced Graphene Oxide for Efficient Organic Dye Removal in Aqueous Environments

Sunith Madduri, Raghava Kommalapati

Ensuring clean water sources is pivotal for sustainable development and the well-being of communities worldwide. This study represents a pioneering effort in the realm of water purification, exploring an innovative approach utilizing modified reduced graphene oxide (rGO) aerogels. These advanced materials hold the promise of revolutionizing environmental remediation efforts, specifically in the removal of organic dyes from aqueous solutions. The focus is on investigating the exceptional adsorption properties of rGO aerogels, enhanced with cysteamine, to comprehend their efficacy in addressing water pollution challenges. The study meticulously delves into the synthesis process of these modified rGO aerogels, employing state-of-the-art techniques to enhance their adsorption capabilities. The characterization methods utilized encompass a wide array of analytical techniques, including X-ray diffraction (XRD) for structural analysis, scanning electron microscopy (SEM) for surface morphology evaluation, Fourier-transform infrared spectroscopy (FTIR) for functional group identification, Thermogravimetric analysis (TGA) for thermal degradation analysis and latent heat capacity were measured by differential scanning calorimetry (DSC). These analyses provide valuable insights into the structural modifications and surface chemistry of the materials. The research comprehensively explores the intricacies of adsorption kinetics, equilibrium, and thermodynamics to unravel the underlying mechanisms governing dye removal. Additionally, the research assesses the aerogels' structural robustness, reusability, and scalability, showcasing their practical feasibility. These findings mark a significant stride in advancing sustainable water treatment methods and pioneering innovative materials with versatile applications in environmental contexts.

139b Carbon Nanofiber-Bridged Carbon Nitride-Fe_2O_3 Photocatalyst: Hydrogen Generation and Degradation of Aqueous Organics

Ashish Bhagat

Graphitic carbon nitride (g-C_3N_4)-supported ferric oxide (Fe_2O_3) nanocomposite modified with carbon nanofibers (CNFs) has been synthesized for the first time for photocatalytic H_2-generation and the degradation of aqueous organics. In its first dual role, Fe_2O_3 served as the photocatalyst and the chemical vapor deposition (CVD) catalyst to grow CNFs over the g-C_3N_4 substrate. Time of CVD was found to be critical for synthesizing Fe_2O_3-CNF/g-C_3N_4 with a good photocatalytic efficiency. The synthesized materials were characterized for various physicochemical and photocatalytic properties. The creation of a Z-scheme heterostructure between g-C_3N_4 and Fe_2O_3 resulted in the greater H_2-generation rate (2095 μmol/g h) than that (1119 μmol/g h) over the substrate. The photocatalytic degradation data showed ~ 95 and 91% removal of methylene blue and 4-nitrophenol in 180 and 300 min, respectively at 22 °C using the Fe_2O_3-CNF/g-C_3N_4 dose of 0.5 mg/mL. The photocatalytic reactions followed the indirect Z-scheme charge transfer path, with the graphitic CNFs acting as the electron-transfer medium between Fe_2O_3 and g-C_3N_4. The simple fabrication route and high photocatalytic efficiency of the ternary CNF-bridged Fe_2O_3-g-C_3N_4 composite system indicate the utility of the material in integrated environmental and energy applications including microbial fuel cells and microelectrolysis cells.

118a Steam Generation: Options for Lowering Carbon Intensity

Jason Garner

Steam is one of the most common sources of process heat in industry. In many industries, and regions, fossil fuel is the primary energy source for generating process steam, which adds to the organization's GHG emissions totals. There are mature and emerging technologies available for lowering or eliminating carbon emissions from the generation of steam. In this presentation, an overview of options for low, or no, carbon emission steam generation will be presented. This presentation will review biomass steam generation, electric boiler technologies, and heat pump systems and will come from the speaker's perspective of past and current projects.

118b Assessing Sustainable Aviation Fuels (SAF) Pathways through the Retrofitting of Gtl Processes: A Step Towards the Integration of Renewable Energy and Ccu Techniques.

Yasmin Abdelkarim, Mohamed Sufiyan Challiwala, Nimir Elbashir

The aviation industry is one of the major contributors to global greenhouse gas (GHG) emissions, which has raised concerns about environmental sustainability and climate change. In fact, it is responsible for over 2% of global CO_2 emissions and contributes to 3.5% of climate change. Due to the mounting pressure to address the environmental impact of the aviation industry, Sustainable Aviation Fuel (SAF) has emerged as a promising solution to neutralize emissions. SAF is derived through two main pathways: Biofuel SAF and Synthetic SAF. Biofuel SAF is made from renewable resources such as energy crops, fat oils, agricultural residues, etc. On the other hand, synthetic SAF is produced using conventional technologies like XTL (where X stands for gas, liquid, or solid, such as natural gas, crude oil, coal, and municipal solid waste). GTL aviation fuel has a carbon intensity value of 100 g CO_2e/MJ, while conventional jet fuel made from crude oil has a slightly lower value of 84 g CO_2e/MJ. However, there are alternative sources, like Synthetic Paraffinic Kerosene (SPK), made from bio SAF, that produce a range of 6 g CO_2e/MJ to 49 g CO_2e/MJ [1], depending on the source and production process. Despite this, most aviation fuel is still produced using synthetic methods, so finding ways to decarbonize these processes is critical. To address this challenge, the present work explores decarbonization solutions by implementing a hybrid Carbon Capture Utilization (CCU) and renewable energy approach in conventional GTL processing facilities.

Our methodology involves the establishment of a reference scenario where a conventional GTL plant is designed for SAF production. Subsequently, we implement a series of modifications: First, we transition the plant's energy source from natural gas to solar energy, reducing its carbon footprint by 65% and fossil fuel reliance. Secondly, we integrate an innovative reforming technology known as CARGEN®, which was previously developed in our research group. This technology enables the recycling of direct CO_2 emissions that is normally flared in conventional plants by reaction with natural gas to produce carbon-neutral or even carbon-negative products; multi-walled carbon nanotubes (MWCNTs) and syngas. Ultimately, we retrofit the reforming process within the GTL plant with renewable methods, including electrolysis and the utilization of the water gas shift reaction for comparison with the developed hybrid approach in the first two processes. After careful analysis and assessment, these changes were compared to the predefined SAF production CO_2 emissions targets in the range of 6 g CO_2e/MJ to 49 g CO_2e/MJ.

Based on our preliminary targeting assessments, we have identified the potential to significantly reduce the carbon intensity of the existing GTL process. Our assessments have shown that we can lower the carbon intensity from 136 g CO_2e/MJ to 13 g CO_2e/MJ by the implementation of solar energy in place of natural gas via an electrification approach. By comparing CCU technologies such as CARGEN®, water electrolysis, or water gas shift reaction; the best options for CO_2 sequestration for achieving net-zero or even carbon-negative emissions are being identified. The comparison is underway to establish both sustainability and economic outlooks of the aforementioned CCU technologies. This research is part of a decarbonization initiative for the process industry, utilizing innovative carbon capture and renewable energy solutions. The presentation will highlight the recent developments within the project, offering insights into sustainability objectives, capital investments, and the economic advantages derived from the strategic transformation of the GTL sector to produce

SAF.

References:

[1] R. G. Grim *et al.*, "Electrifying the production of sustainable aviation fuel: the risks, economics, and environmental benefits of emerging pathways including CO2," *Energy Environ Sci*, vol. 15, no. 11, pp. 4798–4812, Oct. 2022, doi: 10.1039/d2ee02439j.

[2] Z. Ataya, M. Challiwala, G. Ibrahim, H. A. Choudhury, M. M. El-Halwagi, and N. O. Elbashir, "Decarbonizing the Gas-to-Liquid (GTL) Process Using an Advanced 2 Reforming of Methane Process", doi: 10.1021/acsengineeringau.3c00025.

118c Advanced Solutions for Decarbonizing Hydrogen Generation and then the Refineries

Sanjiv Ratan

Based on the current and imminent global drive for tackling climate change, hydrogen is finding its newer evolving and expanded role as an enabler for the imminent energy transition. With the strong push towards 'near-zero' and then to 'net-zero' carbon targets, efforts are gearing up for the projected transforming of the conventional Grey H2 to Blue H2 (with added CO2 capture), with Green H2 (based on water electrolysis using renewable power) expected to catch up in the longer term.

Moreover, hydrogen has been carrying its crucial role in petroleum refining for catering to the increasingly stringent requirements for cleaner and lighter product slate. Accordingly, it has drawn focal attention in terms of advanced hydrogen management as well as optimized on-purpose H2 generation for enhancing overall refinery profitability.

Over the years, the objective targets have progressed from H2 management towards CO2 management, more so since H2 generation itself caries quite a large C-footprint (of 9-12 ton per ton H2). Such efforts however led to only to a marginal reduction in CO2, while not governed by any statuary framework.

However, the "hard-to-abate" industrial sectors like refining and other which involve high temperature processing and requiring fuel firing, are expected to follow the "blue pathways" in support of fossil-continuity with CCS. Based on various continuing technological developments for Post-combustion CO2 capture, mainly to relax its energy and capital intensity as well as to improve process reliability, there is no clear cut and cost-effective process as yet which is also sufficiently deployed commercially. Thus, considering the proposed decarbonization targets of > 85% for a Refinery inherently having multiple fuel-fired heaters, the underlying need for post-combustion CO2 capture makes it quite a challenging and perhaps risky venture on a larger scale.

In addressing the above trends, the paper will describe the innovative and distinctive developments by ZoneFlow Reactor Technologies, firstly in terms of their **ZF structured catalyst for steam reforming** which by replacing the existing pellets, allow a capacity increase of 15-25% combined with enhanced efficiency, thereby also lowering the CO2 per unit H2 both for retrofits as well as new SMRs. And secondly as a bigger breakthrough, a rearranged conventional H2 plant flowsheet called **"Z-H2"** which allows >95% decarbonization through deploying only the well-proven and cost-effective pre-combustion CO2 removal from high pressure clean syngas.

The Z-H2 concept is based on recycling and retaining the total carbon input within the process while having full SMR firing with nearly pure H2. The H2 plant capacity though becoming much larger, also for catering for the H2 fuel requirements for rest of the refinery units replacing the refinery fuel gas, the overall economics remains attractive in the wake of eliminating the post-combustion CO2 capture while also achieving the required refinery-wide net carbon avoidance.

The paper will include comparative techno-economical assessment and NPV analysis for CO2 capture in decarbonizing hydrogen generation using natural gas feedstock in conjunction with utilization of refinery off gases and expanded hydrogen network for the targeted decarbonization of the refinery.

Decarbonizing the hydrogen generation and having centralized pre-combustion capture within the expanded and new H2 plants using Z-H2 process technology for producing extra hydrogen for blending or replacing the hydrocarbon fuels in the refinery heaters will prove to be an effective, proven, and economical route for decarbonization of a refinery. Z-H2 It offers an attractive, easy to execute, reliable and efficient way forward in realizing the imminent and future carbon avoidance mandates.

196a Improved Refinery Processing Strategies

Jate Cox, Hsiang Yee Hoekstra

At SwRI, we have integrated teams working on all facets of renewable petrochemicals and related infrastructure developed for real-world fuel producers. Our expertise ranges over thermochemical methods for conventional and unconventional feedstock cracking and conversions facilitated by scaled-up isomerization and hydrotreating technologies to obtain refined fuels while achieving the desired properties and quality standards. Simultaneously, SwRI also works on engine testing and required adaptations for such renewable fuels. Renewable gasoline and diesel are characteristically like their fossil fuel counterparts and thus can be used as standalone fuels or blended with fossil fuels in internal combustion engines. Another growing field with competent demand is the production of SAF from various sustainable sources. Through numerous and diverse projects completed over decades SwRI promotes an achievable and sustainable energy and material economy. This presentation will cover the different facets of scaling-up and upgrading different feedstocks to renewable petrochemicals while competing and learning from fossil-based petrochemicals.

196b Accurate Prediction of Gasoline Octane Number at Molecular Level

Yuan Zhou

With the increasingly fierce competition in the refining industry, the refineries continue to optimize the process route and adjust the product structure through a series of technologies such as oil upgrading, refining and chemical transformation, molecular management, digital intelligence, and so on, which is expected to maximize the economic benefits of the enterprise. At the same time, the molecular composition in the gasoline pool has brought great changes, making the oil blend face severe challenges.

Traditional oil blending is linear blending based on physical properties, which is difficult to accurately predict octane number due to the ignored nonlinear effect between single hydrocarbon molecules (alkanes, cycloalkanes, olefins, aromatics, etc.) in gasoline pool. Therefore, it is necessary to establish an accurate 'molecular level octane number prediction model', which can realize the blending precisely, save octane number, reduce production cost and provide strong technical support for refineries to improve quality and efficiency.

At present, there are up to 200 to 300 kinds of single hydrocarbon molecules that can be identified in one oil sample, such as FCC gasoline, reformulated gasoline, etherified gasoline, etc., while the number of oil samples available for analysis in a single production cycle is very small, only dozens of kinds. The quantity variance between the two is an order of magnitude. It forms a mass-feature small-sample research problem, which is difficult to predict by conventional model methods. But machine learning algorithms provide solutions.

In this work, 27 oil samples including single hydrocarbon molecular composition data of different carbon numbers were selected, and three models based on machine learning algorithms of BP neural network, support vector machine and random forest respectively were compared to predict octane number. The model with the smallest absolute deviation would be selected as the optimal model. As the results showed that the absolute deviation range of random forest prediction was the smallest, so the random forest model was selected the optimal model. The prediction accuracy of gasoline octane number at molecular level was 1.5 higher than of traditional physical property method.

196d Improving Operational Performance When Processing Opportunity Crudes

Lara Petrishchev

Opportunity crudes have been sought after for several years due to affordability and the challenges include accelerated fouling, bottlenecks, slowdowns, and unscheduled shutdowns. However, a combination of instrumentation and software are continuously improving to allow for crude diversity.

This presentation examines the measurement and control technologies and IIoT strategies that are available to enable refineries to avoid problems and maximize profits while successfully processing opportunity crudes. The technologies discussed allow refiners to quickly characterize crude properties, adjust feedstock blending, monitor for asset deterioration including both corrosion and exchanger fouling and optimize chemical injection and desalter operations effectively. Many of these technologies can be incorporated into an IIoT strategy which would include wirelessHART sensors, remote monitoring, analysis, and interpretation of data, leading to actionable insights that improve operating performance.

Summarizing, it is suggested that moving the perspective from sensor technology to actionable information requirements and digital use of data will provide increased safety combined with enhanced operational profit.

78a Economic Impact and Risk Analysis of Integrating Sustainable Aviation Fuels into Refineries

Nicholas Carlson, Michael Talmadge, Avantika Singh, Ling Tao, Ryan Davis

The growth of the aviation industry coupled with its dependence on energy-dense, liquid fuels has brought sustainable aviation fuel (SAF) research to the forefront of the biofuels community. Petroleum refineries will need to decide how to satisfy the projected increase in jet fuel demand with either capital investments to debottleneck current operations or by integrating bio-blendstocks while aligning with industry decarbonization objectives. Consequently, a Monte-Carlo simulation framework was developed to compare jet production strategies on a risk-adjusted, economic performance basis. At the core of the framework, a refinery optimization model, developed in AspenTech's PIMS linear programming software, was used to identify optimal jet production strategies given crude oil prices and product demands. Randomized crude oil price and product demand projections were then fed to the refinery model while simulating (1) debottlenecking with capital investments and (2) SAF integration scenarios to get distributions of economic performance in terms of net present value. Resulting distribution highlight both the economic impacts and risks associated with both scenarios. Furthermore, incentive structures aiming to de-risk initial SAF production from the refiner's perspective are explored. Results indicate that variable incentive structures with market sensitivity have the potential to greatly diminish the economic risks associated with SAF production and can likely be generalized to de-risk other biofuel investments.

78b Carbon-Negative SAF from Regenerative Agriculture Biomass Via the Alder Renewable Crude Process

Earl Christensen

Alder Renewables' proprietary technology enables the utilization of lignocellulosic biomass to produce hydrocarbon fuels. Fast pyrolysis is a high yield and scalable process that converts abundant biomass into an energy liquid. However, the oil produced contains a high concentration of oxygen and water, imparting low energy density compared to hydrocarbon fuels. In addition, many of the oxygenated species are multi-functional and thermally reactive. Upgrading pyrolysis oil to hydrocarbons using refinery processes has been greatly limited due to poor stability, leading to catalyst fouling and reactor plugging. The Alder Renewable Crude process addresses these barriers by fractionating pyrolysis oil into two streams for upgrading; a stable aromatic fraction for hydroprocessing (ARC) and pyrolytic sugars that has potential to be utilized as a boiler fuel for renewable electricity or converted to biogas (APS). Miscanthus x giganteus (sterile triploid clone) has shown great promise as a pyrolysis feedstock, providing several advantages. This perennial non-invasive, non-food crop requires minimal water or fertilizer and has potential to be grown on marginal lands. Miscanthus produces an extensive root system that reduces erosion and adds nutrients to depleted soil. Moreover, the bulk of the carbon taken up by this plant is sequestered in its root structure. UIUC analyzed field data from two sites and determined that over a 15-year cycle Miscanthus x giganteus had a significantly negative CI score. Alder, in collaboration with INL and BTG, processed one tonne of miscanthus grown by AGrow Tech into pyrolysis oil, which was upgraded to ARC and APS. Working with Honeywell UOP and RPD, the ARC was hydrotreated to produce SAF, which was then analyzed by WSU for critical jet fuel properties. The resulting fuel showed excellent performance on ASTM specification properties, and a detailed life cycle analysis performed by NREL determined the SAF generated maintained a negative CI score. This presentation will provide details on the cultivation, analysis, processing, upgrading, and final fuel properties of carbon negative SAF generated from regenerative agriculture using Alder's ARC process.

78c Maximizing Sustainable Aviation Fuel Production through Optimized Hydroprocessing of Bio-Oils

Kasper Hartvig Lyng Lejre, Magnus Zingler Stummann, Jostein Gabrielsen

Introduction

Achieving climate neutrality in the transportation sectors requires large volumes of truly sustainable bio fuels to provide long-term solutions for sectors where direct electrification is not viable. Biomass can be used as a raw material for the production of fuels with low (potentially zero) carbon footprint [1]. Biofuel processes such as hydrotreating of vegetable oils can convert biomass into fuel through well-established technologies [2]. Topsoe A/S has been developing and refining the HydroFlex™ technology since 2004 to tailor the product and optimize the yields of both renewable diesel (RD) and sustainable aviation fuel (SAF) from bio-oil feedstocks.

Experimental/methodology

In this study, the target was to produce maximum SAF yield while minimizing the loss of overall liquid fuels, SAF+RD, yield. Experiments were carried out in Topsoe's hydrotreating pilot plant facilities, using hydrotreated vegetable oil (HVO) and commercially available hydroprocessing catalyst. The tests were conducted in a once-through, fixed-bed, high-pressure pilot plant setup, loaded with 22 g of TK-920 D-wax™ which is specifically designed for dewaxing HVO and gas oil. The temperature was varied between 280 and 340°C, and the pressure applied was in the range of 40-80 barg. At each condition, gas and liquid samples were drawn and analyzed after reaching line-out. Gas and liquid analyses were carried out in order to close mass balance calculations and giving the basis for calculating yields and hydrogen consumption.

Furthermore, the feed and liquid products were subjected to a range of analytical methods to determine properties such as freezing point, simulated distillation, and density. The liquid products were fractionated to investigate the properties of the SAF and RD fractions separately. More comprehensive methods such as GC×GC-MS/FID/AED and GC-VUV were employed to give in-depth understanding of the chemical reactions taking place over the catalyst.

Results and discussion

Figure 1. The overall SAF+RD yield in vol % FF as a function of the SAF yield in vol % FF.

The HVO was successfully converted to SAF and RD meeting the ASTM D7566 and ASTM D975 specifications, respectively. HVO is a liquid consisting mainly of normal paraffins with carbon numbers in the range C15-C18 resulting in poor cold flow properties. HVO produced by hydrotreating of vegetables oils or animal fats typically has a freezing point in the range of 22-28°C depending on the specific feedstock used. In order to meet SAF specifications both isomerization, to meet freezing point, and hydrocracking, to meet density and boiling point requirements, must be carried out. This is best done in hydroprocessing using a bi-functional catalyst or catalyst system tailor-made to minimize yield loss [3]. In Figure 1, transportation fuel yield, SAF and RD, is plotted as a function of SAF yield. Very high transportation fuel yields are obtained even for SAF yields larger than 20 vol% of fresh feed. The HydroFlex^{TM} process can be applied to advanced bio fuel production where the feedstock is based on e.g. lignocelloulosic- or sewage waste. Recent pilot plant test results show that it is possible to produce SAF based on advanced bio fuel feedstock.

References

[1] McKendry, P. *Bioresour Technol* **2002**, *83*, 37.

[2] Donnis, B. et al. *Top Catal* ***2009**,* *52*, 229.

[3] Kumar, H.; Froment, F. *Ind. Eng. Chem. Res* **2007**, *46*, 4075.

78d Sustainable marine fuel production through mild hydrotreating of catalytic fast pyrolysis oil

Xiaolin Chen, Martha A. Arellano-Treviño, Kellene A. Orton, Abhijit Dutta, Calvin Mukarakate, Kristiina Iisa, Eric Tan

Lignocellulosic biomass-derived biofuels have great potential to reduce sulfur and fossil carbon dioxide (CO_2) emissions from the maritime sector and help achieve the International Maritime Organization (IMO) GHG emission targets and, at the same time, offer an alternative fuel source for meeting the growing demand for marine fuel. Hydrotreating experiments of the heavy organic fraction (bottom layer) of woody biomass-derived catalytic fast pyrolysis (CFP) oil were conducted at 300-350 °C, 103 bar (1500 psi) with weight hourly space velocities (WHSV) of 0.12-0.24 g/(gh) to produce sustainable marine fuel. Analysis conducted following the ISO 8217 demonstrated that our hydrotreated product met selected standard marine fuel properties. The bottom CFP oil with 23 wt% oxygen on a dry basis became significantly deoxygenated with less than 5 wt% oxygen in the hydrotreated oils at all hydrotreating conditions studied. Increasing hydrotreating temperature or decreasing WHSV led to enhanced deoxygenation and hydrogenation, resulting in a product oxygen content below the detection limit (0.05 wt%) at the highest temperature. The hydrotreated oils were distilled at 140 °C to remove water and compounds with low flash points to obtain marine fuel cuts. Hydrotreating at 300 °C with a WHSV of 0.17 g/(gh) resulted in the highest yield of marine fuel fraction with fuel properties meeting all tested ISO 8217 specifications and miscible with very low sulfur fuel oil (VLSFO); this milder condition also leads to savings in hydrotreating costs.

155a Innovative developments in simulation-based, techno-economic and sustainability analysis for biofuels and biochemicals

Michael Talmadge, Abhijit Dutta, Ryan Davis, Ling Tao, Eric Tan, Oscar Rosales Calderon, Sharon Smolinski, Hakan Olcay

NREL's Economic, Sustainability, and Market Analysis (ESMA) team utilizes process simulations for conceptual biorefinery designs using Aspen Plus and integrated in-house tools for plant economic analysis. Scope for process simulation-based analysis includes pathways from agricultural, forestry, algae, and waste resources to fuels and chemicals. In support of Department of Energy goals sustainable aviation fuel (SAF), the ESMA team continues development of resource conversion pathways with focus on both cost and carbon intensity (CI) of SAF and other products. This presentation will review various analysis approaches, scope of conversion pathways, and recent innovations in process simulation-based analysis by the ESMA team such as policy impact analysis and integration with ancillary carbon mitigation technologies.

155b A Refinery Perspective on Decarbonizing with Marine Biofuels

Nicholas Carlson, Michael Talmadge, Eric Tan, Emily Newes, Robert McCormick

This work seeks to understand what biofuel production pathways a refinery might prefer to produce very-low sulfur fuel oil (VLSFO) for marine applications. A comprehensive refinery optimization model was developed using AspenTech's PIMS linear programming software. The model was configured to allow (1) direct blending of soy biodiesel, renewable diesel, Fisher-Tropsch diesel, and several pyrolysis oils and (2) indirect blending of all pyrolysis oils via co-processing in a fluidized catalytic cracker (FCC) and diesel hydrotreater into the marine fuel pool. Results showed preferred pathways to bio-VLSFO production included co-processing low-quality pyrolysis oil in an FCC to blend the resulting biogenic LCO, directly blending soy biodiesel, and directly blending small quantities of pyrolysis oil. Bio-VLSFO production costs were compared with those of fossil-VLSFO subject to different marine fuel demands, benchmark crude oil prices, and biogenic fractions in the finished product. Given benchmark crude oil prices over 60 $/Bbl, bio-VLSFO production appeared to be significantly cheaper than fossil-VLSFO. Corresponding marginal abatement costs of CO_2 mostly ranging from -300 to 350 $/ton-$CO_2$ were also determined using a simplified but novel approach to allow comparison with other decarbonization strategies. This work indicates that low-sulfur contents in biofuels, relatively relaxed specifications for marine fuels, and current difficulties in meeting VLSFO specifications with crude oils can combine to make bio-VLSFO production cost-effective. Moreover, marine fuels appear to be a good entry point for refiners to start decarbonizing with biofuel pathways that could eventually be extended to other product pools.

SPRING24
+20TH GCPS
A Joint AIChE and CCPS Meeting

SAF from Waste and Biomass: Process Simulation and Design of the Second Gasifier and Syngas Cooler for Low Pressure Gasifiers

Martin Stößel
SCHMIDTSCHE SCHACK, ARVOS LLC, USA
11999 Katy Freeway, Suite 260 Houston, TX 77079, USA
martin.stoessel@arvos-group.com

Dr. Mohammad Haghnegahdar, Dr. Maxim Larin, Jens Hetzer
SCHMIDTSCHE SCHACK, ARVOS GmbH, Germany
Ellenbacher Straße 10, 34123 Kassel, Germany
mohammad.haghnegahdar@arvos-group.com, maxim.larin@arvos-group.com,
jens.hetzer@arvos-group.com

Prepared for Presentation at
American Institute of Chemical Engineers
2024 Spring Meeting and 20th Global Congress on Process Safety
New Orleans, LA
March 24-28, 2024

SAF from Waste and Biomass: Process Simulation and Design of the Second Gasifier and Syngas Cooler for Low Pressure Gasifiers

Martin Stößel
SCHMIDTSCHE SCHACK, ARVOS LLC, USA

Dr. Mohammad Haghnegahdar, Dr. Maxim Larin, Jens Hetzer
SCHMIDTSCHE SCHACK, ARVOS GmbH, Germany

Keywords: Gasification, Syngas Cooler, MSW, SAF, Partial oxidation, Simulation, Design

Abstract

The world's first commercial plant to produce sustainable aviation fuel from municipal solid waste began operations in 2022, in the US. The process consists of the main steps. Gasification, 2nd conversion step, syngas cooling, syngas purification and fuel synthesis.

The paper focusses on the 2nd conversion step and the cooling section designed as full heat recovery to improve the overall process efficiency. As the raw syngas from the gasifier still contains some tars and heavy hydrocarbons, a chemical reactor is installed between upstream gasifier and downstream syngas cooler. The main point in the design is to limit the oxygen available to the synthesis gas for the 2nd conversion step. Applying the partial oxidizer reactor (POx) significantly increases the quality of the syngas. The downstream syngas cooler plays a key role in meeting the syngas requirements of the entire process. By cooling the syngas, heat is recovered to produce high pressure steam which improves the overall thermal process efficiency. Downstream of the syngas cooler is the water sump and slag handling unit, where the cooling, collection and transport of ash and slag takes place.

This paper presents the chemical process simulation of this process step as well as an insight into the equipment design.

1 Introduction

Due to growing concerns about global climate issues, there is increasing interest in renewable and alternative energy sources, as well as new technologies with high energy-conversion efficiency. Biomass, which comprises a diverse range of materials including municipal solid waste (MSW), green waste, and agricultural residues, plays a significant role in renewable energy production. As an alternative to fossil fuels, biomass can be converted into various usable forms of energy, such as syngas, biogas, and liquid transportation biofuels[1].

Biomass gasification, including the gasification of MSW, is poised to be a key player in energy production and conversion. Its application holds immense potential for enhancing energy security, mitigating climate change, and promoting sustainable development. Through thermochemical processes, biomass's carbonaceous materials can be transformed into a combustible syngas, with gasification playing a central role.

Gasification occurs when oxygen or air reacts at high temperatures with the available carbon in biomass or other carbonaceous materials within a gasifier. The gasification process involves three interconnected steps: pyrolysis (decomposition), gasification, and partial combustion. Partial combustion provides the necessary heat for the endothermic gasification reactions.

The resulting syngas primarily consists of carbon monoxide (CO) and hydrogen (H_2), both of which find extensive industrial and household applications. Hydrogen serves as a clean energy carrier and has the potential to diversify automotive fuel sources, eventually being produced solely from renewable energy sources. Gasifiers are no longer limited to chemical and petrochemical industries; they now find applications in various other fields as well[2].

2 Overview of a Waste to Jet Fuel Plant with 2nd Conversion Step

The gasification, partial oxidation (POx) as 2nd conversion step and primary gas purification units produce syngas which is processed through the syngas conditioning and purification section before being fed to a Fischer-Tropsch (FT) synthesis unit and product upgrading unit. The main units of a Waste to Jet Fuel plant are shown in Figure 1.

A Waste to Jet Fuel plant typically includes a Feedstock Conditioning Facility at the beginning of the process to refine the input feedstock (MSW or Biomass) into a suitable feed for the subsequent gasification. The plant receives MSW or biomass feedstock via trucks.

The prepared feedstock is then gasified in the Steam Reforming Trains. The cyclone system removes entrained solid particles, including unconverted carbon, ash, and bed material, from the syngas before it is directed downstream to the POx reactor. The Heat Recovery Steam Generation (Syngas Coolers) system recovers heat from the syngas leaving the POx reactor. The cooled syngas from the POx train is then directed to the 1st Syngas Cleaning step, namely the Particulate and Halides Removal train, which includes a Venturi Scrubber and a Scrubbing Packed Tower. Slag degassing and dewatering occur within the Slag Handling Unit.

The syngas stream from the particulate and halide removal train is sent to the next unit, where its pressure is increased. After compression, the remaining contaminants are removed and the hydrogen to carbon monoxide ratio is adjusted via a water gas shift reaction to an optimum composition for the subsequent Fischer-Tropsch (FT) process.

The FT product is a synthetic crude which is sent to the product upgrading unit where it is selectively hydrocracked and hydrotreated before being sent to a fractionation column to recover the primary product, Synthetic Paraffinic Kerosene (SPK), and the by-product naphtha.

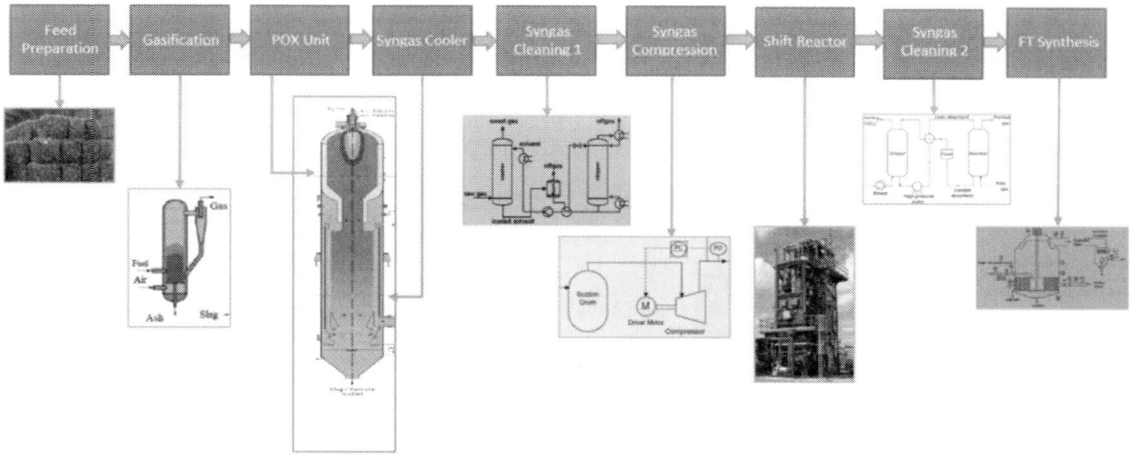

Figure 1. Overview of main units in a typical Waste to Jet Fuel plant

3 The 2nd Conversion step: Schack® Reactor and Syngas Cooler

The raw syngas from the gasifier still contains some tars and heavy hydrocarbons, which becomes sticky when the syngas cools down and thus might attach at the heating surface of the syngas cooler and cause fouling. Since the size of the syngas cooler is proportional to the fouling behavior, it may lead to mammoth equipment size to achieve the required performance. To minimize this risk a second stage gasification based on the approved POx technology has been developed. A further advantage of the 2nd conversion step is that the cracked tars and heavy hydrocarbons improve the overall raw syngas yield. Many process and design issues such as residence time or/and raw syngas inlet into the POx reactor have to be solved. Another challenge is the presence of different contaminants (such as sulfur and chloride) in the syngas, which are contained in the MSW and biomass feed and do not crack during gasification. This defines the material choice and the water treatment process for the water sump. SCHMIDTSCHE SCHACK (SCS) developed a POx based "Tar Cracker" comprising a Radiant Syngas Cooler (RSC) and Convective Syngas Cooler (CSC). The Syngas Cooler Technology is based on the long-time practical experience in the field of special heat exchangers and multiple investigations for coal, crude oil and petcoke gasification processes. Figure 2 presents the schematic diagram of the Schack® POx reactor, RSC and CSC.

The Schack® POx reactor uses a burner to provide the energy required to crack the tars and heavy hydrocarbons. The main point in the design of the tar cracker is to limit the oxygen available to the synthesis gas. This is why it is usually referred to as a partial oxidizer (POx). Applying the POx significantly increases the quality of the syngas. The main function of the POx reactor is to convert the residual hydrocarbons mainly methane and heavy hydrocarbons like tar components into syngas. The POx reactor will also convert some of the ammonia and organic Sulphur in the syngas into N_2/H_2 and H_2S/COS, respectively.

Hot syngas from the POx will then be cooled down in two steps. First in the RSC using radiation as main heat transfer driver, then in the CSC using convection for optimal heat transfer. Downstream RSC heating surfaces a water sump is installed to capture falling down ash / slag particulates from the syngas. Table 1 shows the typical Performance Data of Schack® RSC.

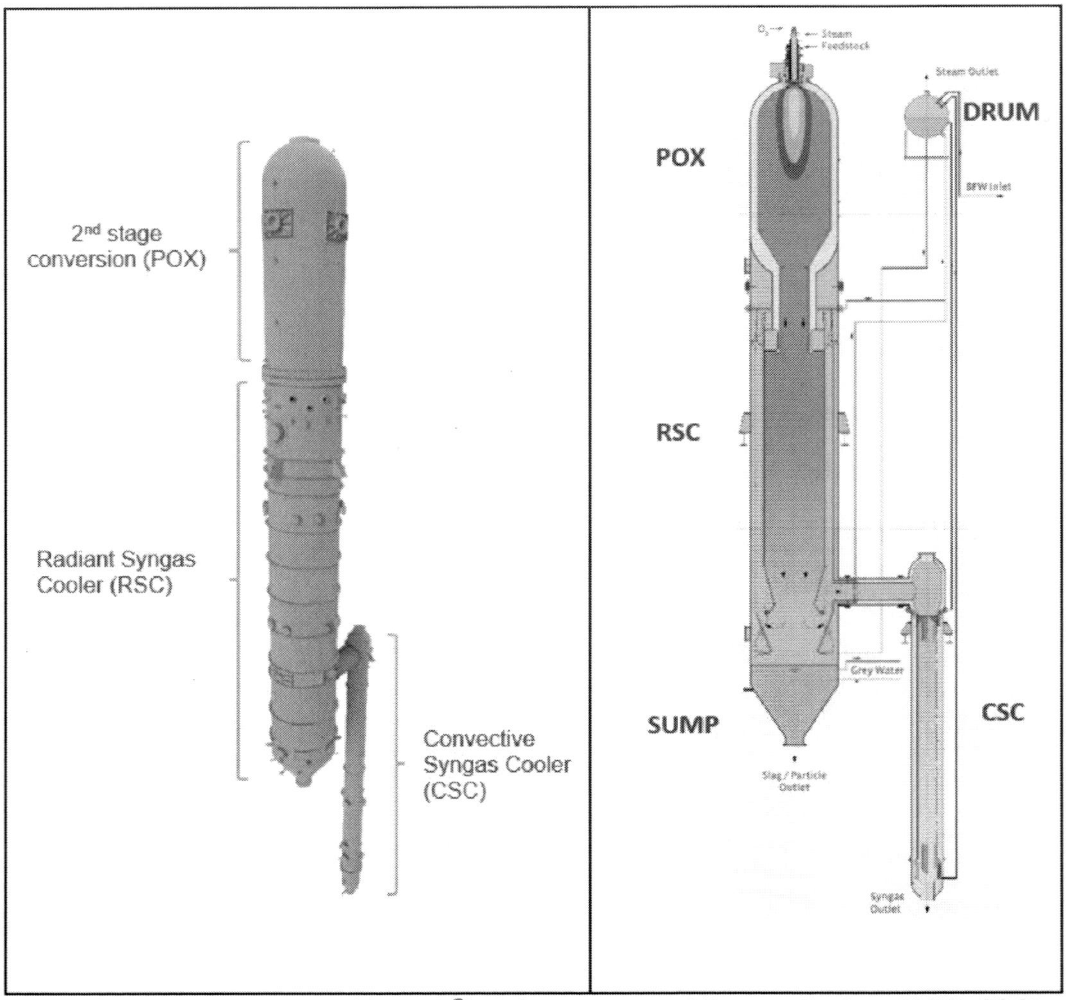

Figure 2. Schematic diagram of Schack®'s POx reactor and Syngas Coolers (RSC, CSC)

Table 1. Typical Performance Data of Schack® RSC

Exchanged Heat	~ 30-40 MW
Water/ Steam pressure	~ 40-70 bar
Steam production	~ 30-70 to/h
Syngas inlet temperature	~ 1200-1500 °C
Dust particles	~ 2-20 g/Nm³
Apparatus length	~ 20-40 m
Apparatus weight	up to 500 ton
Water / Steam Cycle	natural circulation

4 Process modelling

4.1 Modelling of POx Reactor

In the POx there are the following conditions between the syngas components:

- Very good mixing between syngas coming from gasifier, burner fuel and oxidizer
- Sufficient residence time to crack heavy hydrocarbons.

The typical operating conditions in the POx are:

- Pressure from 1.1 bar
- Temperature 1200 – 1500 °C

The POx reactor will convert any residual methane and heavier hydrocarbon components (including tars) of the raw syngas produced in the Gasification Unit into H_2 and CO. Aspen Plus® was used to model the chemical reactions and phase equilibrium in the POx reactor and in the syngas cooler. For the chemical reactions, Gibbs minimization approach was chosen to model the reactions in gas, liquid, and solid phase. Gas phase results were validated by comparing with the pilot plant data provided by SCS partner. Figure 3 shows the simulation results for a MSW feed and confirms the increase of syngas yield in the POx reactor. The amount of H_2+CO increases by 60%pts. and the decrease of Methane and heavy hydrocarbons when having a 2nd conversion step by POx.

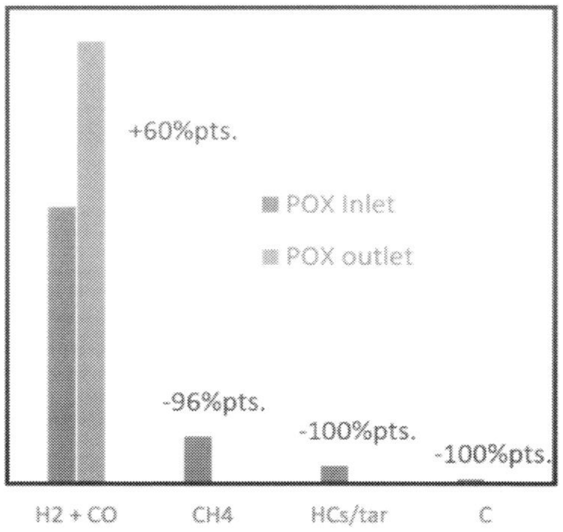

Figure 3. Increase of syngas yield in the POx reactor for a MSW gasification process

Interaction between the trace components containing Cl, S, P and N and ash and particulates was modelled using HSC Chemistry® in equilibrium conditions. Figure 4 shows some selected results regarding the most stable form of each trace component depending on the temperature of the system.

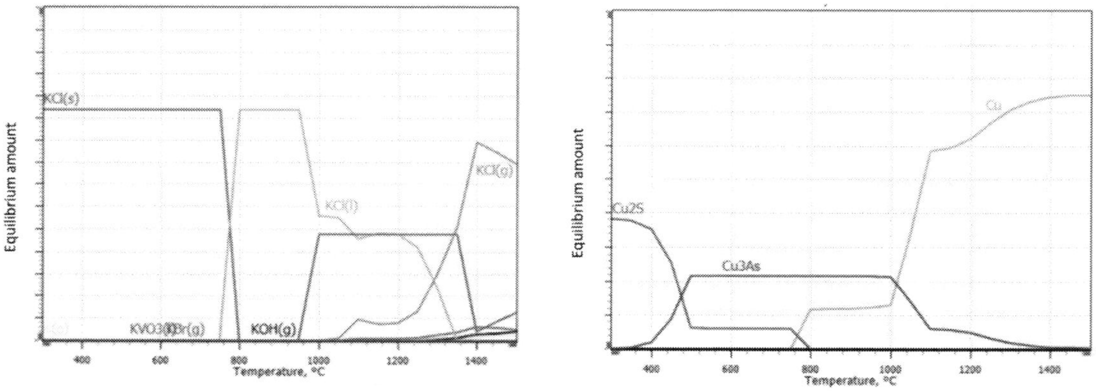

Figure 4. Selected results on the most stable form of trace components (K and Cu) as a function of temperature and syngas composition

The purpose of this simulation is the identification of liquid and solid phases of inorganic species and it's potential impact onto the equipment. Although the simulation does not reflect the reaction kinetics, the results provide a valuable indication.

4.2 Reactions in radiant syngas cooler (RSC)

Since the syngas temperature at POx outlet is high, reaction between the gas components can continue in the cooler. To take this into account, the Syngas cooler is modelled as a chemical reactor. The dominant mechanism of heat transfer is radiation, where in a few meters the gas temperature decreases significantly.

Because of this temperature change, the high temperature syngas leaving the POx which is in equilibrium, starts to react and achieves a new equilibrium. The main reactions are Water Gas Shift reaction (WGS) and steam methane reforming:

$CO + H_2O \leftrightarrow CO2 + H_2$ $\Delta H_{298} = -41.09$ kJ/mol

$CH_4 + H_2O \leftrightarrow CO + 3H_2$ $\Delta H_{298} = 206$ kJ/mol

Since the WGS is exothermic, the decrease of syngas temperature favor's the generation of H_2, which leads to increase of H_2/CO ratio in the syngas stream. Figure 5 presents the results of the simulation regarding the increase of H_2/CO ratio in the RSC for syngas generated from a MSW feed. It can be clearly seen that due to the syngas cooling the H2/CO ratio further improves by 12%pts.

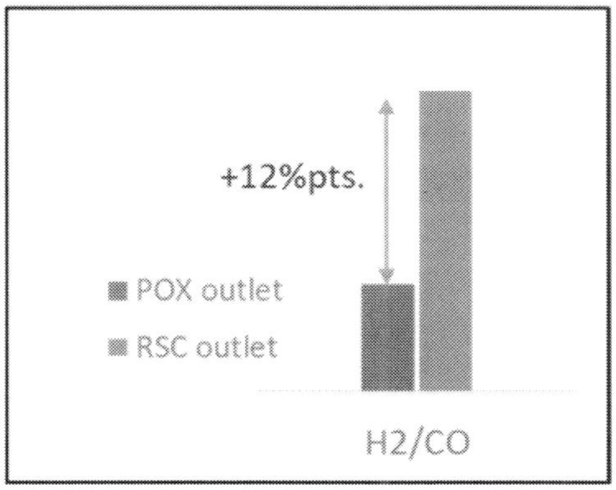

Figure 5. Increase of Syngas yield in the Radiant Syngas Cooler for a MSW gasification process

5 Syngas cooler sump

Some of the particulates and ashes entering the POx reactor with the raw syngas are expected to melt in the POx. However, they start to solidify again when entering the syngas cooler and fall down into the bath water sump at the bottom of the RSC. The water sump will maintain a water level to keep quenching the ash/slag dropping out from the syngas. Water and slag are withdrawn from the bottom of the water sump and are routed to the Slag Handling System, where slag and water are separated. A constant pH-value in the slag bath is maintained using a buffer solution.

As it was mentioned, the main function of RSC Sump is to collect and cool down the large slag and ash particles in the syngas. It consists of a three-phase system, namely Gas (syngas), Liquid (water sump) and Solid (ash/slag).

To model this system the following aspects have been considered:

- Energy transfer from gas to water (radiation and convection)
- Gas and water contacting (mass transfer between gas and liquid)
  The dissolution (equilibrium) of syngas and ash particles into the water were modeled using Aspen Plus® electrolyte package. It is assumed that the water bath consist of two well mixed zones, with the upper zone is in chemical equilibrium with the syngas which leads to the calculation of ion concentrations, salt precipitation and pH in the water bath
- Particles and water contacting/equilibrium
- Temperature distribution in the water bath

A CFD simulation was used to validate the assumptions regarding:

- Gas and solid temperature and velocity profiles
- Ash removal in the sump
- Water evaporation rate from the sump water surface

A typical result of a CFD simulation is shown in Figure 6. The velocity profile above the liquid level shows a small magnitude. Therefore, particles fall into the water sump, which cannot follow the flow turning.

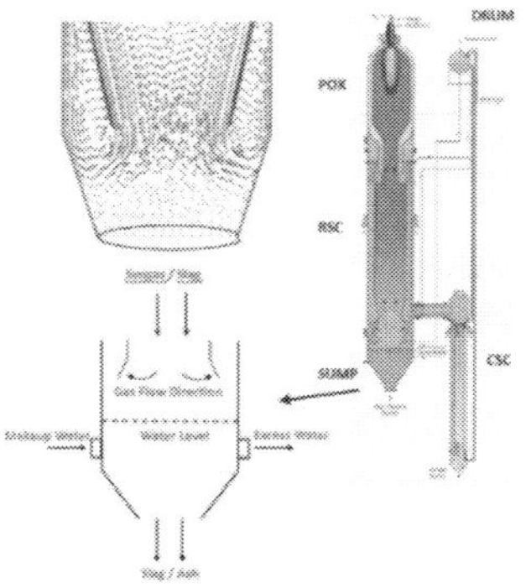

Figure 6. Selected results of a CFD simulation of water bath

6 Summary

The SCHACK® syngas cooler and 2nd stage conversion POx reactor were successfully applied to a commercial renewable gasification plant. It was shown that the POx reactor with its 2nd stage conversion plays a crucial role in improving the quality of the produced syngas in a MSW and biomass gasification processes. By optimizing the syngas composition and applying a full heat recovery for the syngas cooling, this process step with its components contribute to the overall process efficiency and economics.

The POx-RSC system benefits from in-house modelling. This modelling approach helps with the mechanical design and optimization of plant equipment. By fine-tuning of the system parameters, engineers can achieve better performance and reliability.

In summary, these innovations contribute to cleaner, more efficient gasification processes, making progress towards sustainable energy production.

7 References

[1] Avraam DG, Halkides TI, Liguras DK, Bereketidou OA, Goula MA. An experimental and theoretical approach for the biogas steam reforming reaction. *Int J Hydrogen Energ* 2010; **35**: 9818-9827.

[2] Begum S, Rasul MG, Akbar D. A Numerical investigation of municipal solid waste gasification using aspen plus. *Procedia Eng* 2014; **90**: 710-717.

155d Wastewater Process Design and Simulation Using Open-Equation Software

Sahand Iman Shayan, Jochen Steimel

We present our experience developing and validating a library of models for municipal wastewater treatment using AVEVA Process Simulation, an open-equation simulation tool. The library contains the core functionality required to represent a typical activated sludge treatment plant, including nitrification, denitrification, sludge production and settlement, and biogas generation via anaerobic digestion. The library incorporates industry standard ASM (Henze) and ADM (Batstone) models. Simulations can be solved in steady state as well as dynamic modes.

By drawing on other model libraries in the simulation environment it is possible to consider energy recovery from biogas, and economic analysis for sustainability. Due to the open-equation nature of the system, the models can readily be adapted or extended by the user if required.

185a Open-Source Python Seeq Packages for Sysid and Optimization Based PID Tuning

Junho Park, Mohammad R. Babaei, Ashwin N. Venkat, John Hedengren

This study presents new open-source Python packages for system identification (SysID) and PID tuning, developed as Jupyter Notebook add-ons to the Seeq software. Seeq is a web-browser-based commercial platform for data exploration, cleansing, inferential building, and other common data analysis. The new additions, SysID and PID tuning packages, build upon the Seeq Workbench software with advanced data analysis and modeling capabilities, tailored for dynamic model identification and control tuning.

The SysID app has six model identification options including ARX, ARIMAX, FIR, transfer function, subspace, and neural network models. It supports a diverse array of linear and nonlinear models, like linear time-series, state space, nonlinear differential and algebraic equations, continuous and integer variables, and machine-learned models, based on the Python Gekko modeling platform. In addition to the traditional format of models, the SysID app also enables constrained SysID, Transformer Neural Networks, Transfer learning, and Physics-Informed Neural Networks.

In a second contribution, a PID tuning app uses an optimization-based PID retuning process. This method leverages historical setpoint and load disturbance data to refine the PID controller tuning parameters more effectively. The exhaustive search method avoids local minima in an attempt to improve tuning by exploring a predefined PID parameter search space. The reliability of the tuning is enhanced by simulating conditions to be as close as possible to realistic operating conditions, creating a digital twin of the PID controller's characteristics. The presentation includes a discussion of a potential future work to combine the SysID and PID tuning apps with identification to drive more accurate process models from open-loop or closed-loop data for PID retuning.

185b Advanced Strategies for Digitalisation of Petrochemical Processes

Jivathran Saswin

Sustainable industrial development is a key to higher global production, sustained incomes and environment. Digital transformation of chemical processes have significantly drew attention of sustainability players. This paper focuses on how plant digitalization could improve the efficiency, profits and reduce the cost of petrochemical processes.

185c Novel Production Scheduling and Control Integration

Gaya Dandekar

Industrial control and scheduling integration has been studied and new control features are added. Control challenges and advantages are discussed. Applications of artificial neural networks plays a significant role in the evolution of production scheduling. The collaboration between academia and industry with a goal of systematical controls integration would be greatly beneficial.

174a Accelerating the Engineering of Sustainable Processes through the Next Generation of Process Simulation

Ian Boys, Cal Depew

Globally the process industries are facing imperatives to address environmental changes caused by chemical process manufacturing. We are promoting hydrogen as an alternative fuel that reduces greenhouse gas emissions and a circular economy to reduce waste.

New processes must be put in place in the next decades which will require an acceleration of the engineering, design, and commissioning of these new plants to meet government objectives. AVEVA's new Process Simulation platform can help address a digital transformation of process engineering that are difficult to achieve with mainstream single-purpose process simulation tools to address these sustainability imperatives:

1) An open modeling approach to shorten the development time of a new 'renewables' model library for wind, solar, electrolysis, pyrolysis, to model the hydrogen value chain in mainstream process flowsheet simulation and deploy circular processes.

2) An Engineering 'Digital Twin' that accelerates the engineering of the new plants by allowing collaboration of process engineering with other engineering disciplines beyond the process world to verify equipment, piping, and control design through hydraulic analysis and dynamic simulation.

3) An Operations 'Digital Twin' that reuses the engineering process models online to minimize their equivalent greenhouse gas emissions, possibly augmented by data-driven models that use Machine Learning and Artificial Intelligence.

174b *Dynamic Simulation of LNG Fuel Tank Pressure for LNG-Fgss, Building and Using Digital Twin with AWS*

Keisuke Suzuki, Katsuhide Hashimoto, David Hill, Kenichi Matsuoka, Aaron Herrick

1. *Abstract*

LNG is becoming increasingly viable and popular as a marine fuel because it is clean-burning, produces fewer NOX emissions, and has the lowest emissions level per joule of energy released of all fossil fuels.

The use of LNG on board is attractive because it complies with strict emission rules set out by the International Maritime Organization (IMO). Fuel gas supply system (FGSS) for LNG fueled ships is a fuel gas handling system, and LNG fuel is delivered to the engine and boiler system for propulsion and power generation of vessels.

This paper shows application of dynamic simulation to analyze and predict the LNG tank process behaviors (pressure, temperature and LNG aging) for LNG fueled ship, using a digital twin (CHEMCAD simulator) with cloud hosting, and verifying the behavior of controllability of fuel gas supply pressure from FGSS to engine system, and the interactions of BOG compressor control and LNG pump control etc. The dynamic simulation is executed in real time, and the results of calculation are transferred to LNG fueled ship through Sea-Navi® system.

2. *Overview of LNG fuel gas system for LNG fueled ship*

LNG is the cleanest fossil fuel currently on the market. Vessels with a FGSS meet international measures and regulations to reduce emissions and contribute to a cleaner environment for future generations.

The main components of the FGSS are as follows:

- *LNG bunker skid*
- *Cryogenic type C tank*
- *LNG pump (LNGP)*
- *BOG compressor (BOGC)*
- *LNG vaporizer & LNG gas heater for LNGP line*
- *Pre-heater & After-cooler for BOGC line*
- *Buffer tanks*
- *Process control system*

The simplified process flow diagram for FGSS is shown in Fig.2.1.

<Fig.2.1: Simplified process flow diagram for FGSS>

As shown in Fig.2.1, the fuel gas flow rate to the main engine is based on BOGC pressure control and LNGP pressure control.

Herein, we focus on the dynamic process study for LNG fuel tank behavior according to fuel gas flow and pressure to fuel gas consumers supplied by BOGC and LNGP.

The controllability of LNG fuel tank pressure is studied, based on the points of view of the operating period of both controls, the change of operating points. Then, the related optimum control parameters are calculated through the dynamic simulations at the same time.

<Fig.2.2: Simplified data communication between ship and Amazon Web Services (AWS) through Sea Navi® system

Fig.2.2 shows the system architecture of the data communication between LNG fueled ship and AWS. The system conducts the online dynamic simulation using digital twin in AWS cloud system.

174c Process Digital Twins for Optimal Plant Operations

Bill Poe

Process Digital Twins for Optimal Plant Operations

Process simulations have been used in most industries since the invention of computers primarily for engineering efforts in process design, troubleshooting and debottlenecking. More recently process digital twins have emerged as a tool to support plant operations. Petroleum refining and major chemical plants pioneered the use of process digital twins for feedstock selection, planning and process optimization of complex units. As the robustness of process digital twins has continued to improve, their adoption has expanded to other industries and for a wider variety of purposes.

A robust process digital twin has the attributes of 1) an accurate representation of the current performance of an asset over its full range of operation including the disposition of each molecule, and 2) fully automated model updating and runs that complement business workflows. This digital twin can support process engineering, project engineering, planning and scheduling, operations, process control and plant maintenance.

The future process digital twin is expected to merge with other digital twins including behavioral digital twins and supply chain digital twins. Next generation digital twins will also include more artificial intelligence capabilities to model poorly understood processes and enable autonomous operations by detecting and responding appropriately to abnormal conditions.

This presentation will cover:

- the history of process digital twins
- the structure and functions of a robust process digital twin
- the benefits of process digital twins with actual case studies
- the future for process digital twins

174d CFD Modeling of Coke Calcination Rotary Kiln for Troubleshooting and Operation Optimization

Krishnadash S. Kshetrimayum, Minkyu Han, Younghee Lee

In Petroleum industry, Coke Calcination is a process of producing Calcined Petroleum Coke from Green Coke (raw petroleum coke) and the process is carried out in a large energy intensive rotary kiln (> 3m diameter and >60 m length for instance). The Calcined product is sold to aluminum and steel industry where it is used to make carbon anode for their smelting processes. In Coke Calcination Unit of Project A, repeated failure of the Rotary Kiln inner refractory lining and castables were detected which frequently interrupted the initial operating years after commissioning. Determining root case of the failures and adjusting the unit to safe operating condition was a paramount concern. To this end, Computational Fluid Dynamics (CFD) model of the Coke Calcination Rotary Kiln was developed to simulate the kiln operation and get insights of the phenomena occurring inside the kiln. Analysis revealed that excess fine content (more than 20%) in feed made the process sensitive to other process operating variables and when subjected to overly excess combustion air supply and improper control of firing hood pressure, undesired combustion pattern and flame shapes were produced whose repeated occurrence caused the refractory failures. Optimal range of operating variables for stable operation was recommended based on operation optimization.

99a Simulation of Biogas Production Using Rigorous Fermentation Model in Aspen Plus

Shu Wang, Selim Anavi, David Tremblay

This presentation is to demonstrate how to simulate manure to biogas conversion in an anaerobic batch fermentation tank, using biocomponents and the built-in fermentation unit operation in Aspen Plus. The fermentation kinetics are modeled in a single reactor in order to represent wastewater treatment processes involving mixed cultures of micro-organisms. The fermentation model provides sufficient flexibility to closely describe real-world phenomena and capable of accurately predicting gas yield and production rates. It can be used to help process engineers have a deeper understanding and insight for the manure to biogas conversion process, and lead to ideas for process improvement and optimization.

99b CFP Regenerator Model Development

Bruce Adkins

Catalytic Fast Pyrolysis (CFP) is a promising technology for producing fungible transportation fuels from biomass. Just as in petroleum resid or vacuum gas oil (VGO) cracking in FCC units, carbonaceous coke deposits on the catalyst, and the heat from combusting the coke in a regenerator is used to drive the endothermic cracking reactions. However, biogenically derived coke is different from VGO and resid coke, and design of oxidative regenerators and regeneration processes must take these differences into account.

In coordination with the ChemCatBio, Advanced Catalyst Synthesis and Characterization (ACSC) and Catalyst Deactivation Mitigation (CDM) BETO consortia, the Consortium for Computational Science and Chemistry (CCPC) has been investigating the kinetics of coke generated on ZSM-5 based CFP catalysts and using these kinetics to build industrial scale reactor models. Two different ZSM-5 based catalysts have been investigated: a Geldart B (nominal 600-800 mm) coked to relatively high levels (12-15 wt% coke on catalyst), as would occur in a bubbling bed CFP reactor, and a Geldart A (FCC-type, nominal 80 mm) catalyst coked to much lower levels (< 2% coke on catalyst), as would occur in an FCC-type CFP reactor operating at typical catalyst circulation rates.

Kinetics model development was based on Temperature Programmed Oxidation (TPO) data generated at Pacific Northwest National Lab (PNNL), complimented by advanced microscopic characterization at Oak Ridge National Lab (ORNL). Fixed bed finite element computational models were used to assess mass and heat transfer limitations in the TPO experiments and incorporate learnings from characterization such as intrapellet coke deposition patterns. The fixed bed models were also used to test an ensemble of kinetic hypotheses including direct surface oxidation of C and H to form CO, CO_2 and H_2O (Eley-Rideal mechanisms) and homogeneous gas phase oxidation of CO to CO_2.

From the ensemble of models, the ones best fitting the experimental data were down selected, incorporated into the MFiX software suite and used to simulate a 10 ton per day (TPD) demonstration-scale regenerator. Parametric studies of different gas-solids drag closures, air flow rates, solids inventories etc. were conducted and the process was optimized. Furthermore, an industrial-size catalyst regenerator for a 2500 TPD scale CFP unit was simulated with MFiX-Exa, a true exascale capable code, producing the first (to our knowledge) exascale regenerator simulation. With the MFiX-Exa model, different configurations of air grids and various operational modes were studied. The study has high value for addressing the potential problem of "afterburn" (CO combustion in freeboard) common to many FCC units.

AUTHOR INDEX

Transport and Energy 2021

Held at the 2021 AIChE Annual Meeting

Boston, Massachusetts, USA and Online
7 - 11 November and 15 - 19 November 2021

AIChE
The Global Home of Chemical Engineers

Transport and Energy 2021

Held at the 2021 AIChE Annual Meeting

Boston, Massachusetts, USA and Online
7 - 11 November and 15 - 19 November 2021

AIChE
The Global Home of Chemical Engineers